Biomes of the Past and the Future

Karen Donnelly

The Rosen Publishing Group's
PowerKids Press™
New York

To my family, David, Cathy, and Colleen

Published in 2003 by The Rosen Publishing Group, Inc.
29 East 21st Street, New York, NY 10010

First Edition

Editor: Gillian C. Brown
Book Design: Michael J. Caroleo

Photo Credits: Cover, title page, back cover, p. 7 © Digital Vision; all border images © Weatherstock; pp. 4, 19 © Digital Stock; pp. 8, 12 © CORBIS; p. 11 © Tom Bean/CORBIS; p. 15 by Michael J. Caroleo; p. 16 © EyeWire; p. 20 © Gary Braasch/CORBIS.

Donnelly, Karen J.
Biomes of the past and the future / by Karen Donnelly.— 1st ed.
p. cm. — (Earth's changing weather and climate)
Includes bibliographical references (p.).
ISBN 0-8239-6215-6 (lib. bdg.)
1. Biotic communities—Juvenile literature. 2. Nature—Effect of human beings on—Juvenile literature. [1. Biotic communities.] I. Title.
QH541.14 .D66 2003
577.8'2—dc21

2001005543

Manufactured in the United States of America

Contents

Biomes

A biome is an area of Earth with a special kind of **climate**. Earth has seven different kinds of biomes. These biomes are rain forest, desert, grassland, **temperate** forest, **taiga**, **tundra**, and **aquatic**. Animals and plants in each biome have **adapted** to their **environment**. For example, the **arctic** fox's thick, white, winter fur hides it from the prey that it hunts. This thick fur also keeps it warm. If the climate in a biome changes, plants or animals must adapt. This means they change the way they live, or, over generations, they develop specialized features called **adaptations**. If changes in the climate happen too suddenly, animals or plants might not be able to adapt quickly enough. They might become **extinct** if the climate becomes too hot or too cold.

The arctic fox has two layers of fur to keep it warm. This is an adaptation to the cold environment where it lives.

Rain Forest

The rain forest biome can be tropical, or hot, or it can be temperate. Tropical rain forests have very hot and wet climates with more than 100 inches (254 cm) of rain per year. They are home to as many as 40 percent of all plant and animal species on Earth. Tropical rain forests are on or near the **equator**. At one time, tropical rain forests covered almost 14 percent of Earth's surface. Today they cover only 7 percent of Earth's surface! The trees have been cut down for the lumber, and the land has been cleared for farming.

Temperate rain forests are usually found in mountainous areas. One can be found in the Olympic Mountains of Washington. This rain forest receives more than 200 inches (508 cm) of rain per year.

Tropical rain forests can be found in South America, Asia, and Africa. Millions of different animals live in a rain forest.

Desert

Desert biomes are very dry regions with little rainfall and very few plants. About 20 percent of Earth's surface is covered by desert. Desert lands might be increasing because of **global warming**, which is caused by natural and human-made events.

There are two different kinds of deserts. Hot deserts, such as the Mojave, can be found in the southwestern United States. North Africa's Sahara is another hot desert. Animals, such as coyotes, have adapted to desert life. They hunt in the cool of night. Desert plants have also adapted. They use little water.

There are also cold deserts. They are not hot, just dry. The Gobi Desert in Northern China is a cold desert.

In the hot desert, temperatures can reach up to 129 °F (54°C). In the cold desert, temperatures get up to only 70°F (21°C).

Grasslands

Grassland biomes get their name from the grasses that cover the landscape. The average rainfall is between 10 and 20 inches a year (25–51cm). This makes it difficult for most types of trees to grow in the grassland biome. Grasslands cover 21 percent of Earth's land area.

Tropical grasslands, called savannas, are hot, with from 20 to 50 inches (51–127 cm) of rain per year. They are found in Africa, Australia, South America, and India. The rain falls from 6 to 8 months during the year. Several months of **drought** follow. **Seasonal** fires often break out. The fires keep large trees and shrubs from growing. The dry grass burns easily but grows again when the rains return.

There are fewer grasslands today because of farming and **overgrazing** by cattle.

This picture shows grasslands near Red Hills and Thunder Basin National Grassland.

Temperate Forest

Temperate **deciduous** forests have trees with broad leaves. Maple, oak, and sycamore trees grow in deciduous forests.

In fall the leaves of deciduous trees change color and drop off. Without green leaves, **photosynthesis** shuts down, and the trees stop growing. This adaptation prevents damage to trees during cold weather and helps trees to stay alive during winter. Warm weather and spring rains help the buds of new leaves to form. The average temperature of a temperate forest is about 50°F (10°C).

Temperate forests are found in North America, Europe, and East Asia. Manhattan Island, home of New York City's skyscrapers, was covered in temperate forest centuries ago.

The temperate forest is home to many animals, such as bears, deer, squirrels, raccoons, skunks, birds, and reptiles.

Taiga

The taiga biome is found in the northern areas of North America, Europe, and Asia. Winters in the taiga are long and dark. For much of the year, it is very cold with a lot of ice and snow. Low temperatures in the winter can fall to -60°F (-51°C).

Most trees in taiga forests are **coniferous** evergreens with dark green, needle-shaped leaves that stay green year-round. Taiga regions have late springs and short summers. The dark green color of coniferous trees helps them to hold heat from the sun and start photosynthesis when spring begins. Small animals, such as red squirrels, voles, wolverines, lynxes, and pine martins, are at home in the taiga.

In the summer, thousands of birds nest in the taiga, because there are a lot of insects to eat.

Taiga summers can be 70°F (21°C). There are days during the summer when the sun never sets, because the taiga is so far north.

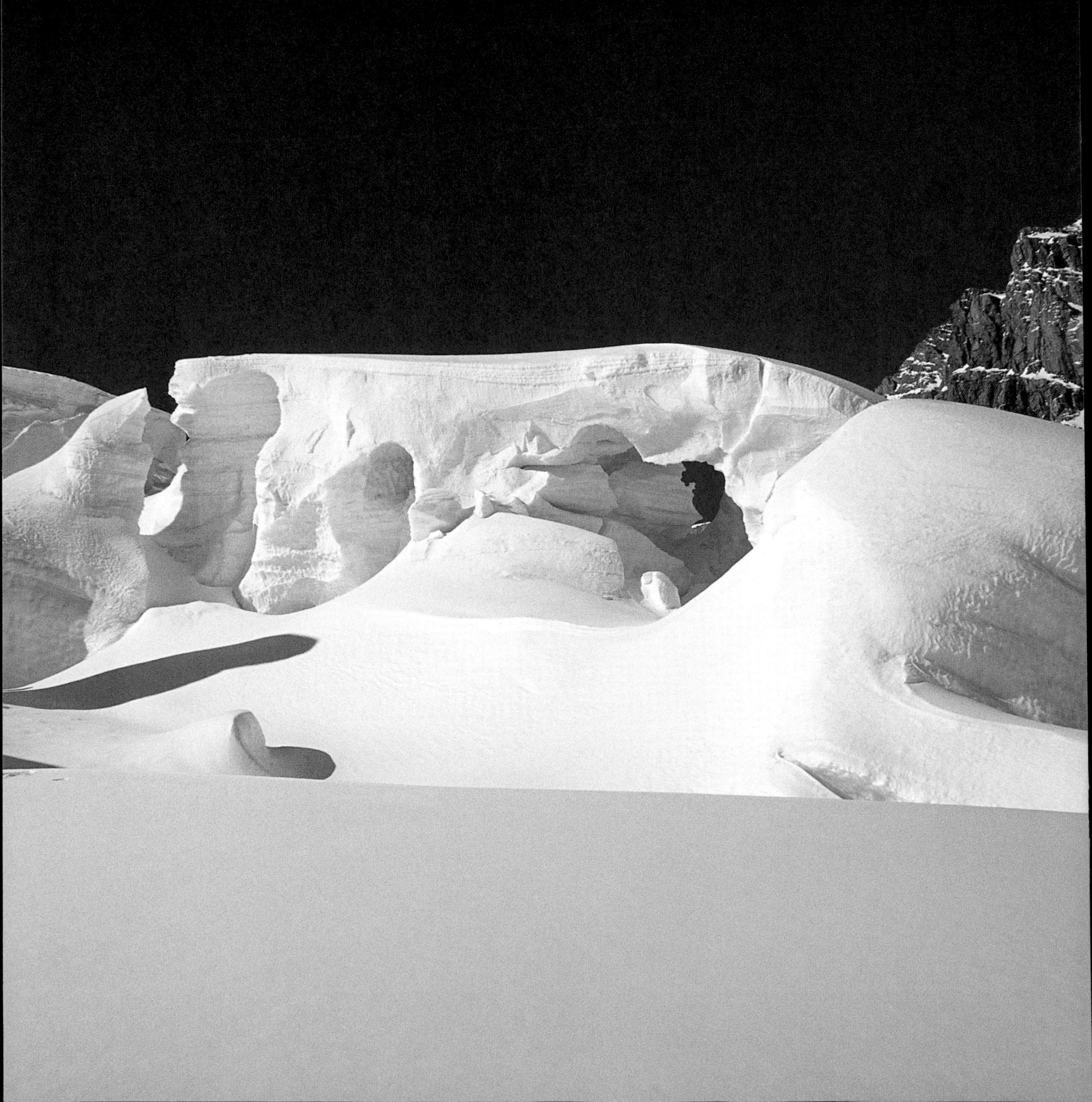

Tundra

Tundra is a Russian word meaning "treeless land." The tundra is the coldest biome on Earth. The arctic tundras are at the northernmost part of Earth. They surround the North Pole. Alpine tundras are found on high mountaintops. Tundras have long, dark, cold winters with temperatures that can fall to -60°F (-51°C). Summer temperatures average only as warm as from 32°F to 50°F (0°–10°C).

Animals, large and small, live in the tundra biome. Arctic foxes, polar bears, caribou, mosquitoes, trout, cod, ravens, and falcons are all able to survive the freezing temperatures and the low rainfall found in the arctic tundra. These animals have adapted to breed and to raise their young very quickly, during the tundra's short summers.

The ground in the tundra, which is shown here in winter, never completely thaws. The layer of permanently frozen ground is called permafrost.

Aquatic

Water covers nearly 75 percent of Earth's surface. Freshwater aquatic biomes include ponds, lakes, and streams. Ducks nest and frogs lay eggs among the grasses along the banks of freshwater biomes. Saltwater biomes include the world's oceans and areas where streams and rivers join the ocean. Coral reefs, such as the Great Barrier Reef off the coast of Australia, are saltwater aquatic biomes. Marine algae, tiny plants that grow on the surface of the ocean, give off much of the world's **oxygen** while taking in large amounts of **carbon dioxide**. Around the world, the plants and the animals that live in freshwater and saltwater biomes are in great danger from pollution.

Animals of all kinds make the aquatic biome their home, including lobsters, sea lions, seagulls, sea turtles, and salmon.

Humans and Biomes

People have had a large effect on the world's biomes over time. Thousands of years ago, forests in Europe were cleared to make way for farms, villages, highways, and cities. Every day thousands of acres (ha) of Amazon rain forest, in South America, are being destroyed to make room for farms and ranches. Farmers in the midwestern United States planted crops. As they did, herds of bison and pronghorn deer lost their grazing lands and disappeared. In the West, gray wolves were driven nearly to extinction by cattle ranchers who wanted to protect their cattle. From tropical rain forests to the tundra, biomes are being affected by the actions of humans.

This forest in Costa Rica was cleared and is an example of the effect that humans have on land.

Changing Biomes

When changes happen slowly, animals and plants can adapt. For example, birds find new places to nest when the climate changes in their biome. The birds might carry seeds to their new homes, and plants may grow with the right temperature and rainfall. These adaptations can occur only if they happen slowly.

In the past, natural causes created change in the biomes, such as the Pleistocene Ice Age.

Humans are changing the environments in Earth's biomes very quickly. Scientists believe that rising temperatures, called global warming, may affect the world's climate. The long-term effect of human activity on the world's biomes will be closely watched by scientists.

Glossary

adaptations (a-dap-TAY-shunz) Changes an animal or a plant makes over time, like a polar bear's thick fur that keeps it warm.
adapted (uh-DAP-tid) Changed to fit new conditions.
aquatic (uh-KWA-tik) Having to do with water.
arctic (ARK-tik) The area around the North Pole.
carbon dioxide (KAR-bin dy-OK-syd) A gas that plants take from air and use to make food.
climate (KLY-mit) Average weather conditions over a long period of time.
coniferous (kah-NIH-fur-uhs) Trees that have needlelike leaves and grow cones.
deciduous (deh-SIH-joo-us) A type of tree that loses its leaves in the fall.
drought (DROWT) A period of time with little or no rain.
environment (en-VY-urn-ment) All the living things and conditions of a place.
equator (ih-KWAY-tur) An imaginary line around Earth that separates it into two parts, northern and southern.
extinct (ik-STINKT) To no longer exist.
global warming (GLOH-buhl WARM-ing) A gradual increase in the temperature on Earth.
overgrazing (oh-ver-GRAYZ-ing) When animals eat most of the plants in an area.
oxygen (AHK-sih-jin) A gas in air that has no color, taste, or odor and is necessary for people and animals to breathe.
photosynthesis (foh-toh-SIN-thuh-sis) The process in which green plants make their own food from sunlight, water, and carbon dioxide.
seasonal (SEE-zun-ul) Happens every year during a particular season.
taiga (TY-guh) Forests with conifer trees that start where tundras end.
temperate (TEM-peh-ret) Not too hot and not too cold.
tundra (TUHN-druh) A cold, treeless plain with permanently frozen soil.

Index

Web Sites

To learn more about biomes, check out these Web sites:
www.panda.org/kids/wildlife
www.worldbiomes.com

DATE DUE
